2nd Natural Gas Utilization Workshop 2018

College Station, Texas, USA
12 - 14 August 2018

ISBN: 978-1-5108-7282-0

Printed from e-media with permission by:

Curran Associates, Inc.
57 Morehouse Lane
Red Hook, NY 12571

Some format issues inherent in the e-media version may also appear in this print version.

Printed by Curran Associates, Inc. (2018)

For permission requests, please contact AIChE
at the address below.

AIChE
120 Wall Street, FL 23
New York, NY 10005-4020

Phone: (800) 242-4363
Fax: (203) 775-5177

www.aiche.org

Additional copies of this publication are available from:

Curran Associates, Inc.
57 Morehouse Lane
Red Hook, NY 12571 USA
Phone: 845-758-0400
Fax: 845-758-2633
Email: curran@proceedings.com
Web: www.proceedings.com

TABLE OF CONTENTS

GENERAL SUBMISSIONS

ADVANCES IN CATALYST/PROCESS CHEMISTRIES FOR NATURAL GAS CONVERSION

ADVANCES IN REACTION ENGINEERING AND SEPARATIONS FOR NATURAL GAS CONVERSION

PROCESS SYSTEMS ENGINEERING FOR NATURAL GAS SYSTEMS

POSTER SESSION

A Blow of Fresh Air to the Oxidation of Light Alkanes

Commercially, 95% of the hydrogen in US is produced by steam-methane reforming of natural gas. However, SMR converts the carbon to CO_2, and limited by endothermic reaction and energy-intensive pressure-swing H_2 separation. Catalytic decomposition of methane (CDM) has been extensively researched at the laboratory. However, the loss of catalytic reactivity, system pressure buildup, and lacking of CO_2-free external heat supply greatly hurdles the industrial deployment and economic feasibility of CDM process.

Since 2013, our team has strived to develop an innovative modular process for coproducing value-added carbon nanotubes and CO_2-free H_2 by nature gas upgrading.***Protected by >3 patents and >10 journal publications, we have successfully integrated auto-thermal pilot-scale chemical looping combustion for CO_2-free heat supply, >95% H_2purity production, >90% carbon separation efficiency in advanced annular carbon stripper.***

In these patented innovations, the Cu-Fe/Al_2O_3 (or Co-Ni/SiO_2-Al_2O_3) was initially reduced exothermally with methane, and reduced oxygen carrier was directly used as catalyst for CDM to produce hydrogen. Depending on chemical composition of catalyst formulation, the tip-growth or base-growth CNTs formed during the methane decomposition was separated by annular separator (patented). In last stage, the reduced catalyst, after CNT harvesting, was exothermally regenerated by air, and recirculated back to reducer for next reaction cycle. Taking advantage of auto-thermal chemical looping operation, this process enables CO_2-free heat supply for endothermic methane decomposition. The same CuO-Fe_2O_3 material functioned as oxygen carrier for chemical looping and as a catalyst for methane decomposition. Unique annular configuration in fluidization separator greatly improve carbon separation with lower solid feed rates. The preliminary TEA suggested significant improvement in energy efficiency (45% savings on electricity), and greater than 50% capital cost reduction compared with state-of-the-art technologies for H_2 production due to elimination of a number of unit operations: WGS, PSA and CO_2 capture units, which seamlessly aligned with RAPIDâ€™s missions.

Coking and Sintering Resistance of ALD Catalyst for Dry Reforming of Methane

Authors:

Afzal, S., Texas A&M University
Prakash, A., Texas A&M University at Qatar
Elbashir, N., Texas A&M University at Qatar

Atomic Layer Deposition (ALD) is a technique to coat thin films in the order of nanometers on solid substrates. It has been shown that the selectivity/activity of catalysts can be altered by the proper tuning of these overcoats[1].

In this work, the objective is to investigate the effect of ALD on the performance of a commercial 20% Ni/Al_2O_3 catalyst in dry reforming conditions. Using TMA as precursor, various cycles of alumina are overcoated on the commercial catalyst. This catalyst is tested against the uncoated catalyst at different temperatures. Apart from stability testing, the crystallite size of the nickel nanoparticles is determined after definite time periods. The crystallite size is calculated from the dispersion which is obtained from H_2 chemisorption. The ALD coating on the nickel active material is purported to prevent sintering by anchoring the nickel nanoparticles. The amount of coke deposited after each run is also estimated by a Temperature-Programmed-Oxidation (TPO) done at the end of each run. At lower temperatures (~600 Â°C), coking is supposed to be the dominant deactivation mechanism and at higher temperatures (~800 Â°C), it is sintering. By studying both these phenomena at different temperatures for ALD with different number of cycles, this work will help better understand the role of the overcoat in the performance of ALD catalysts, and help identify the operating conditions where ALD catalysts have a positive effect on catalytic performance.

REFERENCES

(1) O, B. J.; K Jackson, D. H.; Lee, J.; Canlas, C.; Stair, P. C.; Marshall, C. L.; Elam, J. W.; Kuech, T. F.; Dumesic, J. A.; Huber, G. W. Catalyst Design with Atomic Layer Deposition.*ACS Catal.* **2015**, *5*, 1804â^'1825.

Energy and Economic Metrics for Evaluating Chemical Processing Technologies Associated with Oxidative Dehydrogenation- Ethylene Case Study

Authors:

Gaffney, A., Idaho National Laboratory

An economic and environmental approach for analyzing the potential of a new chemical processing technology for multiple grades of ethylene is described. In order to optimize design and implementation of new processes, a set of performance metrics are established and calculated for light olefins production. The performance metrics are those that could be used by the chemical industry in analyzing the potential of new technologies. Economic evaluations are based on the Cost of Production in relation to current market pricing (spot or contract). The Required Netback is a methodology practiced in industry to define instantaneous production economics. The RNB represents the sum of the Variable Cost, Fixed Cost and Capital Recovery all expressed as currency per mass unit of the main product, ethylene. This is then compared to current pricing; if the RNB is less than the market price then the technology will have favorable economics and increased profit margins. This approach uses current state-of-the-art technology as a baseline for evaluating the promise of a novel technology. In the case of ethylene via oxidative dehydrogenation, the improvement in Cost of Production comes from a drastic reduction in capital cost, energy requirements, and carbon footprint. In this current analysis, oxidative dehydrogenation of ethane is compared with Conventional Steam Pyrolysis and is shown to be superior in all performance metrics.

Enhanced Methane Conversion Via Improved Dry Reforming Catalysts and on-Line Optimization

Authors:

Praserthdam, S., Texas A&M University
Balbuena, P. B., Texas A&M University

Besides the direct use of methane from natural gas to generate electricity, the conversion via the dry reforming process (DRR) to produce syngas is one of the cost-effective routes for natural gas utilization. Here we report an improved optimization component which can be applied to an onâ€'line optimization in the DRR. This optimization approach is demonstrated as a two-step procedure: catalyst identification and optimization. The ratings concept proposed as the DRR computational catalyst evaluation tool was employed to identify the catalystâ€™s reactivity and stability. Each catalyst is given unique reactivity and stability ratings as (RTâ€'S, RTâ€'R) indexes based on its activation energy obtained from density functional theory. These indexes characterize the catalyst ability to (1) activate reactants (CH_4 and CO_2) and (2) form and remove coke compared to the reference catalyst. The indexes are located on the reactivity and stability surface to predict DRR and coking rates at chosen operating conditions. The coking boundary proposed in the concept is used to determine cokeâ€'resistant DRR catalysts, where to meet the acceptable criteria catalysts indexes must be in the cokeâ€'removal zone possessing a higher rate of coke removal than formation. The extended ratings concept expands its application to incorporate the experimental apparent activation energy to abovementioned indexes. Next step is to generate the database for DRR and coke formation/removal rates of a chosen catalyst at different operating temperatures and CO_2/CH_4 feed ratios. This information is incorporated into the onâ€'line optimization. The goal is to achieve the conditions where for given changes in methane concentration from the source (e.g., from flue gas) the reaction rate in the reactor can be sustained.

References

[1] Iyer et al. Ind.Eng.Chem.Res, 56 (2017) 8622-8648.

[2] Zhang et al. Comput.Chem.Eng, 19 (1995) 305-310.

[3] Praserthdam et al.Cat.Today, 312 (2018) 23-34.

[4] Praserthdam et al. React.Kinet.Mech.Cat., 122 (2017) 53-68.

Methane to Value Added Chemicals Using a Solid Superacid Catalyst

Authors:

Spivey, J. J., Louisiana State University
Kanitkar, S., Louisiana State University
Carter, J. H., Cardiff University
Hutchings, G. J., Cardiff Catalysis Institute
Ding, K., Northwestern University

The shale gas revolution has had a significant impact on the production of natural gas and its conversion into higher-value products. The most commonly practiced methane conversion process involves the use of a co-reactant as an oxidant (typically H_2O). Non-oxidative routes involve the use of what is known as a superacid, a compound that has acidity higher than 100% sulfuric acid. A number of superacids are liquids; e.g., HF-SbF_5 [1], FSO_3H-SbF_5 [1]. Recently, gas-phase HBr-$AlBr_3$ [2] has shown superacidity for the conversion of methane into value added chemicals, but neither liquid- nor gas-phase superacids can serve as the basis of a practical process because separation of the superacid from the products is unrealistic. The literature has reported attempts to prepare solid superacids using sulfated zirconia[3], but both poor yields and rapid deactivation have limited their further use. The challenge here is to show that the superacidity that has been demonstrated in gas-phase catalysts can be shown in a solid.

Here, we propose a new solid catalyst based on $AlBr_3$ supported on zeolites. DRIFTS using pyridine showed presence of an additional Brønsted acid site as compared to the H-ZSM5 base zeolite. NH_3-TPD quantitatively confirmed the presence of more acid sites than base zeolite. This increase in acid site density and creation of a superacid site was able to catalyze methane into higher-value added chemicals such as C_2-C_8 hydrocarbons. It appears that the reaction is produced in a step-chain oligo-condensation mechanism. Additional tests on ethane oligomerization showed a similar range of hydrocarbon products, and with a higher TOF, as expected.

1. G. A. Olah, G. Klopman and R. H. Schlosberg, *J. Amer. Chem. Soc.*, 1969, **91**, 3261-3268.
2. S. Vasireddy, S. Ganguly, J. Sauer, W. Cook and J. J. Spivey, *Chem. Comm.*, 2011, **47**, 785-787.
3. D. Fraenkel, *Catal. Letters*, 1999, **58**, 123-125.

Two New Gtl Processes from Gti

Authors:

Marker, T., Gas Technology Institute

GTI is currently working to develop two new processes for the conversion of natural gas to liquids. Both will be done in low cost modular systems so that they are economic at relatively small scale.

The first process is called Cool GTL. Cool GTL can be used to convert high CO2 content natural gas, biogas streams, and gas produced from gas plants to liquids. In Cool GTL we convert high CO2 feeds to synthesis gas at a 2/1 H2/CO ratio using a novel CO2/steam reforming catalyst which has high stability. The reformer catalyst has been tested for 500 hours and shows no deactivation. The product from the reformer then goes directly to a low temperature bubbling fluid bed Fischer Tropsch reactor which makes hydrocarbon liquids from the synthesis gas. A unique catalyst formulation enables the fluid bed to make no wax. The fluid bed allows excellent temperature control. Cool GTL is also under development and still requires longer integrated testing.

The second process is called the SoftOx Process. In this process sour (high H2S content) natural gas is converted to high sulfur liquids and clean gas. The high sulfur liquids will be sent down the pipeline with the crude oil and the clean gas will be burned to generate electricity. This project is at an early stage but has shown very promising results with as much as 26 wt. % sour liquids generated per pass from sour natural gas feeds. A unique catalyst was discovered for the process which minimizes coking. Integrated long term testing is need to move forward in this development. The goal of this process is to monetize sour natural gas and prevent sour gas flaring.

Valorization of Natural Gas Liquids from Shale Gas

Authors:

Ribeiro, F. H., Purdue University

The last 10 years have witnessed an unprecedented revolution in the domestic supply of shale hydrocarbon reserves with the potential to lower the cost of our nationâ€™s electricity, power, chemicals, and fuels for the next 100 years. If new technologies can be developed, energy experts project the U.S. will transition from a net importer to net exporter of energy resources by 2030. In addition, substitution of natural gas for coal for electrical power generation has, and will continue, to be a major contributor to the reduction in U.S. greenhouse gas until economic, renewable energy future can be developed. We will present a vision on how to create a transformative engineered system to convert light hydrocarbons from shale resources to chemicals and transportation fuels in smaller, modular, local, and highly networked processing plants.

Electrochemical Membrane Reactors for Combined CO2/O2 Capture from Flue Gas and in Situ Conversion of Natural Gas

Authors:

Huang, K., university of South Carolina

Abstract

This presentation summarizes recent development at presenterâ€™s lab on high-temperature mixed conducting membranes for carbon/oxygen capture and reactors for in situ utilization of the captured CO_2/O_2 as a soft oxidizer to catalytically convert methane into syngas through dry and dry-oxy reforming, into ethylene through oxidative coupling, and ethane (in NGLs) into ethylene through oxidative dehydrogenation. The working principles of the new electrochemical membranes will be first explained in detail. Then extensive laboratory results will be presented for each type of conversion with catalysis mechanisms explained. Finally, the potential of the new membrane reactors will be discussed.

Intensified Fluidization Process for Industrial-Scale Coproduction of Value-Added Carbon and CO2-Free H2 from Natural Gas

Authors:

Tian, H., West Virginia University
Hu, J., West Virginia University

Commercially, 95% of the hydrogen in US is produced by steam-methane reforming of natural gas. However, SMR converts the carbon to CO_2, and limited by endothermic reaction and energy-intensive pressure-swing H_2 separation. Catalytic decomposition of methane (CDM) has been extensively researched at the laboratory. However, the loss of catalytic reactivity, system pressure buildup, and lacking of CO_2-free external heat supply greatly hurdles the industrial deployment and economic feasibility of CDM process.

Since 2013, our team has strived to develop an innovative modular process for coproducing value-added carbon nanotubes and CO_2-free H_2 by nature gas upgrading.***Protected by >3 patents and >10 journal publications, we have successfully integrated auto-thermal pilot-scale chemical looping combustion for CO_2-free heat supply, >95% H_2purity production, >90% carbon separation efficiency in advanced annular carbon stripper.***

In these patented innovations, the Cu-Fe/Al_2O_3 (or Co-Ni/SiO_2-Al_2O_3) was initially reduced exothermally with methane, and reduced oxygen carrier was directly used as catalyst for CDM to produce hydrogen. Depending on chemical composition of catalyst formulation, the tip-growth or base-growth CNTs formed during the methane decomposition was separated by annular separator (patented). In last stage, the reduced catalyst, after CNT harvesting, was exothermally regenerated by air, and recirculated back to reducer for next reaction cycle. Taking advantage of auto-thermal chemical looping operation, this process enables CO_2-free heat supply for endothermic methane decomposition. The same CuO-Fe_2O_3 material functioned as oxygen carrier for chemical looping and as a catalyst for methane decomposition. Unique annular configuration in fluidization separator greatly improve carbon separation with lower solid feed rates. The preliminary TEA suggested significant improvement in energy efficiency (45% savings on electricity), and greater than 50% capital cost reduction compared with state-of-the-art technologies for H_2 production due to elimination of a number of unit operations: WGS, PSA and CO_2 capture units, which seamlessly aligned with RAPIDâ€™s missions.

Methane Pyrolysis on Complex Melts: Reactive Separation for Hydrogen and Carbon

Authors:

McFarland, E. W., University of California, Santa Barbara
Metiu, H., University of California - Santa Barbara
Gordon, M., University of California Santa Barbara
Palmer, C., University of California, Santa Barbara
Zeng, J., UCSB

Utilization of low cost natural gas may be limited someday by societyâ€™s desire to reduce carbon dioxide emissions. The value of the resource as a fuel and source of hydrogen may be preserved by either capturing and storing the carbon oxides or avoiding their production altogether. Methane pyrolysis has long been utilized as a means of producing hydrogen from methane. If high value carbon products are desired for the carbon black market, plasma-based systems show promise to produce low pressure hydrogen and valuable carbon black. The high production costs can be supported by the value of the carbon product. Lower costs of production are possible using previously demonstrated molten metal-based processes whereby high-pressure hydrogen may be produced together with separable solid carbon which may be sold until the markets become saturated then readily and permanently disposed of. Results of investigations of methane pyrolysis in multiphase reacting systems containing molten metals and/or molten salts will be presented. Fundamentals of C-H bond activation on melt surfaces will be discussed for metals and salts and distinctions made as to how the subsequent reaction pathways can produce different types of carbons ranging from graphitic to disordered. Specific compositions of molten metal alloys and mixed molten salts will be discussed which have particularly high activities for pyrolysis and produce different types of carbon products. The practical problem of producing a commercially acceptable carbon product will be discussed along with the heat transfer and materials of construction challenges.

Mixed-Oxide Redox Catalysts for Shale Gas Valorization Under Cyclic Redox Schemes

Authors:

Li, F., North Carolina State University

The significant increase in shale gas production has generated substantial economic incentive to convert light alkanes, e.g. methane and ethane, to value-added products. This is especially the case in distributed locations where methane and ethane are often subjected to flaring or rejection due to production fluctuations and limitations in transportation capacities. Although there are well-established commercial processes for methane and ethane conversion, they tend to be energy intensive with significant carbon dioxide emissions. The large energy consumptions for these processes are often resulted from the high endothermicity of methane/ethane conversion reactions, the complexity in product separations, and/or the energy consumptions for cryogenic air separation. The complexity of these conventional processes also limit their applications to convert distributed shale gas resources.

In this presentation, we discuss the chemical looping schemes as potentially attractive approaches for shale gas valorization, particularly in distributed, modular applications. The chemical looping process utilizes an oxygen carrier, also known as redox catalysts, to convert shale gas components via an indirect reduction-oxidation (redox) scheme: the redox catalyst first releases its lattice oxygen to convert methane or ethane into value added products such as syngas (via methane partial oxidation) or ethylene (via ethane oxidative dehydrogenation); the oxygen depleted redox catalyst is then regenerated by reacting with a gaseous oxidant, e.g. air, steam, or CO_2. A few examples of utilizing the chemical looping concept for methane partial oxidation, CO_2 splitting, ethane oxidative dehydrogenation, and light alkane liquefaction are provided. Strategies to effectively design and optimize redox catalysts are also discussed.

Reaction Engineering and Separations Challenges for Natural Gas Utilization: An Industry Perspective

Authors:

Powell, J. B., Shell International Exploration & Production

While mega-scale conversion of methane to liquid products has been commercialized, wide scale use of methane for production of liquid fuels and petrochemicals has not occurred, despite recent price advantages for gas relative to petroleum. Limited yields of targeted C_2^+ products from C_1 (methane) is a challenge which may require advances in reaction engineering and integrated separations, to augment traditional catalysis research for both oxidative and reductive pathways for methane utilization. Distributed production from stranded gas sites has also lagged relative to plants based on economics afforded via centralized production from major gas fields. Simplified, lower capital intensity processes with coupled reaction and separation and standardized modular designs may enable distributed gas processing to be competitive. This seminar will present an industry perspective on the challenges of methane conversion to targeted products, and opportunities for reaction engineering and process intensification to improve economics for methane utilization.

The impact of modular chemical process intensification (MCPI) on distributed gas conversion

Authors:

Bielenberg, J., RAPID

The rise in natural gas production and the shift in the demand slate for fuels & chemical products has create a significant potential opportunity for new technologies in gas conversion. However, there exist significant challenges in the space of gas conversion including (1) high capital costs that drive projects to very large scale to achieve improved capital efficiencies, (2) significant thermodynamic/kinetic challenges in existing gas conversion technologies, and (3) significant downstream separation costs. Process intensification and its application to modular processing has the potential to address all three of these hurdles.

RAPID is a public-private institute that is focused on catalyzing change in the process industries through the use of MCPI. In this talk we will look at the case for MCPI as applied to gas conversion, we will lay out key gaps in the gas conversion space that have been identified in the area of gas conversion, and we will give a short review of 4 projects that RAPID is currently funding in the area of gas conversion.

Zoneflow Structured Catalytic Reactors for Steam Methane Reforming: Experimental Studies of the Reaction Kinetics and Fluid Dynamics and Multi-Scale Modelling

Authors:

De Wilde, J., Université Catholique de Louvain (UCL)
Minette, F., Universite catholique de Louvain (UCL)
Ratan, S., ZoneFlow Reactor Technologies LLC

The multi-scale modelling of ZoneFlow structured catalytic reactors for steam methane reforming (zoneflowtech.com) is addressed. Advantages offered by ZoneFlow reactors compared to conventional packed beds include reduced pressure drop, improved heat transfer and increased catalyst effectiveness. Aspects of the catalyst, kinetic modelling, reactor modelling and required scale-bridging strategies are discussed.

The ASC (Alloy Surfaces, Co. Inc.) deformable Ni catalyst that was used is a thin, nanostructured and adherent Ni coating inter-diffused into a metal foil substrate that undergoes an alloying heat treatment followed by a chemical treatment. The kinetic modelling experiments were carried out in a micro-packed bed reactor set-up. The reactor design and operating conditions ensure plug flow, isothermal operation, small pressure drop and absence of interfacial and intra-catalyst transport limitations. Regression followed by physico-chemical and statistical testing were applied for the kinetic modelling and parameter estimation. The rate determining step-based Hougen-Watson Langmuir-Hinschelwood approach was applied to derive rate expressions that account for the details of the reaction mechanism and facilitate use in and integration of the reactor model. Based on possible reaction mechanisms and rate determining steps, 30 different sets of rate expressions were tested.

To account for the complex fluid dynamics, a CFD reactor model was developed. The RANS approach and standard k-epsilon model with wall functions were applied, introducing the turbulent viscosity, conductivity and diffusivity to account for the influence of turbulence. The turbulence model parameters were independently determined from cold flow pressure drop tests in a wide range of flow rates. Radiative heat transfer and thermal conduction in the reactor tube and in the reactor internals coated with catalyst were taken into account. The Rosseland-WSGGM was used to describe radiative heat transfer. The catalyst effectiveness factors were calculated from the intra-catalyst diffusion-reaction equations and the independently determined reaction kinetics was coupled with the CFD code.

A General Framework for the Evaluation of Direct Nonoxidative Methane Conversion Strategies

Authors:

Maravelias, C., University of Wisconsin-Madison
Huang, K., University of Wisconsin-Madison

We study single-step natural gas conversion technologies that directly convert methane to olefins and higher hydrocarbons. Despite the relative simplicity of these technologies, the development of processes based on these approaches remains challenging. Accordingly, we utilize process synthesis, modeling, and optimization to assess the economic feasibility of direct nonoxidative methane conversion strategies. We develop a flexible approach that allows for the systematic evaluation of various technology alternatives and for the identification of the key technology gaps that must be overcome. The results of our analyses demonstrate that an economically feasible direct methane conversion process is contingent upon fundamental research advances in the area of catalytic conversion to increase methane conversion to hydrocarbon products (e.g., coke formation less than 20% and a minimum conversion to products of 25%). Upon this development, further efforts can be devoted to improve ethylene selectivity as well as reduce catalyst cost and overall capital costs.

Developing Ways to Harness Natural Gas in Africa

It has large incentive to produce more clean liquid fuels from increased natural gas supply. However, current syngas and the Fischer-Tropsch (FT) reactor technologies have low efficiency and high cost. Steam reforming (SMR) is widely used for syngas production. Its carbon efficiency is less than 50%. Autothermal reforming (ATR) has carbon efficiency about 70-74% but it uses pure methane only. It is very expensive to build a gas plant to separate the C2+ from methane. The temperature at the top of the ATR reactor is as high as 2200 °C, which makes the reactor fabrication and maintenance become very expensive.

Dason Technology developed a short contact time catalytic partial oxidation (SCOR) technology. With the SCOR technology, the following catalytic reactions are carried out in one reactor vessel at very short residence time (~ms):

$CH_4 + 0.5O_2 = CO + 2H_2$

Dasonâ€™s SCOR reactor operates under moderate temperature and pressures, and has low cost to fabricate and operate. >90% C-atomic conversion efficiency was demonstrated for conversion of natural gas to syngas.

Molecule Works Inc. invented a compact monolith reactor technology for conversion of syngas to liquid hydrocarbons. The monolith reactor is packaged with monolithic catalyst blocks comprising a number of small, straight flow channels with active catalysts as the channel wall. The monolith structure substantially enhances gas/liquid/catalyst contacting efficiency and mass transfer rates, and provides 3-dimensional, continuous solid matrix for thermal conduction. Greater than 90% one-pass syngas conversion is achieved in a 20-cm long monolith reactor tube at <10% selectivity to methane. The monolith reactor allows the syngas conversion be conducted effectively at relatively low pressures (~10bar) and tolerates presence of CO_2 and N_2.

Combination of those two new reactor technologies significantly simplifies separation, reaction, and heat transfer process steps, and makes it possible to develop a compact GTL plant.

Modular and Agile Design for Unconventional Natural Gas Utilization

Authors:

Hasan, M. M. F., Texas A&M University
Arora, A., Texas A&M University
Li, J., Texas A&M University

With increasing volatility in chemical market, fixed *stick-built* plants are finding it challenging to meet evolving uncertainty and variability in product demands [1]. The chemical plants of the future need to be more modular and flexible in terms of raw material feedstocks utilized, intermittent energy supplied, commodity chemicals produced, and production capacities achieved [2]. The challenges faced in increasing plants flexibility can be addressed through modular manufacturing wherein off-the-shelf operational modules are manufactured independently and are interconnected in a plug-and-play fashion. The use of modular equipment manufacturing and operation has several advantages including specialized equipment manufacturing; lower labor, logistics and manufacturing costs; faster deployment and time-to-market; and favorable economies of mass production [3]. Specifically, such modular production facilities are well-suited for production of specialty chemicals with small capacities or high volatility.

To this end, we develop a unified design and optimization framework for (i) mass-producing modular processing units with fixed design specifications which can be used in diverse chemical industries with inherent synergies; and (ii) developing flexible chemical processes which are suitable for toll-manufacturing and produce a variety of value-added chemicals using different feedstocks. The versatility of the framework will be demonstrated for producing several value-added chemicals from natural gas using flexible chemical process with modular processing units.

[1] T. Bieringer, S. Buchholz, N. Kockmann, Future Production Concepts in the Chemical Industry: Modular--Small-Scale--Continuous, Chem. Eng. Technol. 36 (2013) 900â€“910.

[2] R.E. Swaney, I.E. Grossmann, An index for operational flexibility in chemical process design. Part I: Formulation and theory, AIChE J. 31 (1985) 621â€“630.

[3] M. Baldea, T.F. Edgar, B.L. Stanley, A.A. Kiss, Modular Manufacturing Processes: Status, Challenges and Opportunities, AIChE J. (2017).

Multi-Scale Energy Systems Engineering for Optimal Natural Gas Utilization

Authors:

Pistikopoulos, E. N., Texas A&M Energy Institute, Texas A&M University
Demirhan, C. D., Texas A&M University
Tso, W. W., Texas A&M University

Multi-scale energy systems engineering is a methodologic and generic framework to address complex energy and environmental problems via a holistic and system-based approach and arrive at realistic integrated solutions. This framework is used to tackle the problems existing in design, control, and operation of energy systems and their supply chains in an integrated manner. Optimal solutions are provided to the decision-makers for systems ranging from nanoscale, micro-scale, mesoscale to mega-scale levels over horizons that range from milliseconds to months or year.

Methodologies in energy systems engineering focusing on optimizing natural gas utilization include, superstructure optimization with high-fidelity and data-driven modeling, mixed-integer linear or nonlinear programming, global optimization, optimal design and control under uncertainty, and life-cycle assessment. Aforementioned concepts and methods are illustrated by presenting their applications in a set of various real-life examples, such as (i) modeling of newly emerging natural gas conversion technologies like microchannel or chemical looping reactors [1], (ii) optimal flowsheet design of ammonia-methanol coproduction from natural gas by using process synthesis technique with simultaneous heat, power, and water integration [2] (iii) supply chain optimization of natural gas to liquid transportation fuels (GTL) for nationwide, regional, and state-wide scales [3].

References

1. Tso, WW; Niziolek, AM; Onel, O; Demirhan, CD; Floudas, CA; Pistikopoulos, EN. Enhancing Natural Gas-to-Liquids (GTL) Processes Through Chemical Looping for Syngas Production: Process Synthesis and Global Optimization. Computers & Chemical Engineering 2018, 113, 222-239
2. Tso, WW; Demirhan, CD; Powell, JB; Pistikopoulos, EN. Toward Optimal Synthesis of Renewable Ammonia and Methanol Processes (RAMP). 13th International Symposium on Process Systems Engineering (PSE 2018) 2018; Accepted manuscript.
3. Elia, JA; Baliban, RC; Floudas, CA. Nationwide, Regional, and Statewide Energy Supply Chain Optimization for Natural Gas to Liquid Transportation Fuel (GTL) Systems. Industrial & Engineering Chemistry Research 2014, 53 (13), 5366-5397.

Natural Gas Hydrate Formation Prediction and Prevention Strategies in Gas Transmission and Processing

Authors:

Shu, B., Certified Technologies
Hou, X., The Chinese University of Hong Kong

Objectives

One of the most critical flow assurance challenges during oil and gas exploration and production is hydrate formation, which can stop the flow and even totally block all production. HYSYS and ProMax are the major software using for hydrate formation prediction. This project is conducted by utilizing the software to predict gas hydrate formation conditions, and to develop hydrate formation strategies based on the simulation data.

Methods

The K-value method, Peng-Robison state of equation in HYSYS V8.6 and ProMax V3.2, and experimental method are applied for hydrate formation prediction in natural gas pipeline systems.

Results

3.1 CO_2/H_2S fraction

The acidic gases (H_2S, CO_2) that are part of hydrate composition and can affect the hydrates formation conditions. We can see that higher concentration of CO_2 in the gas stream can lower the hydrates forming temperature, while H_2S will cause the opposite effect.

3.2 Wellhead

HYSYS and ProMax were used to predict Lavan-3 gas well and Salman wellhead hydrate formation conditions, and their prediction errors were under 5% compared with experimental data.

3.3 Thermodynamic inhibitors

The methanol/Ethane glycol has been injected at 0% 10%, and 20% of weight at the aqueous phase. Then the hydrates formation temperature and pressure are simulated were compared with the field pipeline data and experimental data.

3.4 TEG dehydration

50MMSCFD of natural gas has been treated at TEG plant, at a circulation of 8 GPM of lean TEG. The water content in the gas stream has removed from 75.90 lbm H2O/MMSCF to 4.74 lbm H2O/MMSCF. The hydrates forming temperature has suppressed from 52.66 °F to 23.47 °F.

Applications

The formation of hydrate can cause significant losses in oil and gas production processing and transmission procedure. Results from HYSYS and ProMax were used here to predict wellhead or pipeline hydrate formation conditions, thermodynamic inhibitor injection rate, and TEG circulation rate.

Optimization Approach to Reduction of CO2 Emissions for Syngas Production Involving Dry Reforming

Authors:

Afzal, S., Texas A&M University
Sengupta, D., Texas A&M University
El-Halwagi, M., Texas A&M University
Elbashir, N., Texas A&M University at Qatar

Dry Reforming of Methane (DRM) is a widely studied reaction which uses CO_2 and CH_4 to produce synthesis gas ($CO+H_2$) with $H_2/CO \sim 1$. Since the process uses a waste stream (CO_2), this reaction system has received considerable attention from industry and academia like. However, there are still significant challenges in the path of its commercialization. This work addresses the life-cycle analysis viewpoint looking at the overall CO_2 balance of the DRM process[1]. All major sources of CO_2 emissions as well as credits are accounted for based on available greenhouse gas emissions database from GREET. The reformer modeling was based on Gibbs Free Energy Minimization Principle[2]. It was found that for the production of high H_2/CO ratio syngas, the DRM process helps in reduction in overall CO_2 emissions and operating costs only when coupled with a post-DRM processing step where the excess CO is removed. Without this additional step (which will add to the capital costs), the DRM process produces a low-quality synthesis gas ($H_2/CO \sim 1$). This synthesis gas has limited utility and hence the DRM process cannot compete with existing established processes of syngas production like Steam Methane Reforming where H_2/CO ratios are above 3. The estimated reduction in terms of CO_2 emissions on an industrial scale is presented for the DRM coupled with post-CO removal.

REFERENCES

(1) Afzal, S.; Sengupta, D.; Sarkar, A.; El-Halwagi, M. M.; Elbashir, N. O. Optimization Approach to the Reduction of CO2 Emissions for Syngas Production Involving Dry Reforming. *ACS Sustain. Chem. Eng.* 2018, *6*, 7532â^'7544.

(2) Noureldin, M. M. B.; Elbashir, N. O.; El-Halwagi, M. M. Optimization and Selection of Reforming Approaches for Syngas Generation from Natural/Shale Gas. *Ind. Eng. Chem. Res.* 2014, *53* (5), 1841â€"1855.

A Novel Annular Microchannel Reactor (AMR) for Small-Scale and Distributed Hydrogen Generation

Authors:

Wilhite, B. A., Texas A & M University
Bossard, P., Power and Energy Inc

This presentation gives an overview of the development of a patented AMR technology for small-scale and distributed hydrogen generation. This technology is based upon a tube-in-tube design that represents an effective miniaturization of the annular packed-bed reformer, such that individual annulus dimensions are on the order of 100 â€“ 500micron. This miniaturization provides order-of-magnitude breakthroughs in rates of heat-transfer to the reforming catalyst for equivalent breakthroughs in gas hourly space velocities; Power+Energy, Inc.â€™s manufacturing facility enables rapid assembly of large (100 â€“ 2,000) networks of parallel tube-in-tube assemblies in a single compact module. We will review proof-of-concept single-AMR experiments, resulting single- and multi-AMR CFD simulations, economic analysis of the AMR technology for replacing conventional methane steam reformers in stranded natural gas utilization, and finally module size and cost estimations based upon experimental data and CFD simulations to-date.

Analysis of Particle and Reactor-Scale Transport-Kinetic Interactions of CO2 Methanation

Authors:

Nanduri, A., Texas A&M University-Kingsville
Mills, P. L., Texas A&M University-Kingsville
C R, P. R., Texas A&M University-Kingsville

The design of an externally cooled fixed-bed reactor for CO_2 methanation has been performed using microscopic forms of reaction engineering models that account for both particle-scale and reactor-scale transport-kinetic interactions. This is an important reaction for the clean manufacture of CH_4. The predictions of various multicomponent species flux models were compared to analyze the transport-kinetics interactions for a spherical catalyst particle shape. Flux models that included Knudsen diffusion resulted in an increase in both the CO_2 conversion and CH_4 yield as the pore diameter was reduced from 40 nm to 10 nm, which suggests the development of novel nanoporous catalytic materials. Both 1-dimensional (1-D) and 2-dimensional (2-D) models for different catalyst shapes were also simulated to quantify particle-scale performance for CO_2 methanation. A spherical catalyst shape was shown to produce a higher CH_4 average concentration, while a novel 4-hole cylindrical shape has the lowest diffusion limitation. The performance of an externally cooled 1-D heterogeneous fixed bed reactor was also simulated using spherical catalyst pellets. It is shown that the high exothermicity of the CO_2 methanation reaction creates significant hot spot formation. In order to control the axial reactor temperature within a user-prescribed bound, the optimal catalyst loading profile was determined by using an energy conservation equation-based approach. It is shown that the reactor hot spot temperature can be controlled within a reasonable range, which provides the basis for development of practical reactor designs for this important reaction.

Compact Modular Reactor Unit for Conversion of Shale/Natural Gas into Liquid Products

Authors:

Xu, B., Dason Technology Inc

It has large incentive to produce more clean liquid fuels from increased natural gas supply. However, current syngas and the Fischer-Tropsch (FT) reactor technologies have low efficiency and high cost. Steam reforming (SMR) is widely used for syngas production. Its carbon efficiency is less than 50%. Autothermal reforming (ATR) has carbon efficiency about 70-74% but it uses pure methane only. It is very expensive to build a gas plant to separate the C2+ from methane. The temperature at the top of the ATR reactor is as high as 2200 °C, which makes the reactor fabrication and maintenance become very expensive.

Dason Technology developed a short contact time catalytic partial oxidation (SCOR) technology. With the SCOR technology, the following catalytic reactions are carried out in one reactor vessel at very short residence time (~ms):

$CH_4 + 0.5O_2 = CO + 2H_2$

Dasonâ€™s SCOR reactor operates under moderate temperature and pressures, and has low cost to fabricate and operate. >90% C-atomic conversion efficiency was demonstrated for conversion of natural gas to syngas.

Molecule Works Inc. invented a compact monolith reactor technology for conversion of syngas to liquid hydrocarbons. The monolith reactor is packaged with monolithic catalyst blocks comprising a number of small, straight flow channels with active catalysts as the channel wall. The monolith structure substantially enhances gas/liquid/catalyst contacting efficiency and mass transfer rates, and provides 3-dimensional, continuous solid matrix for thermal conduction. Greater than 90% one-pass syngas conversion is achieved in a 20-cm long monolith reactor tube at <10% selectivity to methane. The monolith reactor allows the syngas conversion be conducted effectively at relatively low pressures (~10bar) and tolerates presence of CO_2 and N_2.

Combination of those two new reactor technologies significantly simplifies separation, reaction, and heat transfer process steps, and makes it possible to develop a compact GTL plant.

Design and Intensification of Methanol Synthesis from Natural Gas

Authors:

Arora, A., Texas A&M University
Iyer, S. S., Texas A&M University
Bajaj, I., Texas A&M University
Hasan, M. M. F., Texas A&M University

Methanol is a promising alternate fuel to depleting oil and gas reserves and is one of the most crucial chemicals due to clean-burning, energy-storage medium, as a CO_2 utilization alternative, and as a precursor for synthesizing several hydrocarbons. The current methanol production process from natural gas is one of the largest energy consumers in chemical industry due to inherent equilibrium limitations. The process inefficiencies include low single-pass conversions, high recycle ratios, high energy consumption and equipment cost, and catalyst deactivation. The objective of this work is to counter existing methanol process inefficiencies by enhancing single-pass methanol reaction conversions and yields. We design periodic sorption-enhanced methanol synthesis (SE-MeOH) processes for methanol production from natural gas. In SE-MeOH systems, *in situ* adsorption of reaction byproduct(s) favorably shifts the equilibrium towards enhanced methanol synthesis in accordance with the Le Chatelierâ€™s principle.

To this end, firstly, we produce the intermediate synthesis gas from natural gas with desired feed conditions and specifications using a steam methane reforming (SMR) unit. The synthesis gas is then converted to methanol via SE-MeOH process. For capturing the dynamics of both SMR and SE-MeOH processes, we use a generalized adsorption-reaction modeling and simulation (GRAMS) platform, which has been extensively validated with experimental data for both SMR and SE-MeOH cases. GRAMS is coupled with a simulation-based grey-box optimizer for optimizing SE-MeOH process cycle configuration, design parameters and operating conditions. In comparison with base-case industrial methanol reactor, the results indicated 8.17% higher methanol yield and 8.26% lower raw material consumption at a competitive price with a slight compromise on methanol production capacity. Furthermore, results indicated that methanol yield as high as 80% can be obtained with optimal synthesis gas feed composition and flow rate. The developed SE-MeOH processes have smaller carbon footprint, enhanced product quality, and smaller condenser and recompressor size.

Effects of Thermal Activation Conditions on Physicochemical Properties of Nanosheet-Derived Mg-Al Mixed Oxides

Authors:

Ogino, I., Graduate School of Engineering, Hokkaido University
Tanaka, R., Hokkaido University
Mukai, S. R., Graduate School of Engineering, Hokkaido University

Mg-Al mixed oxides possess acid-base sites, and have been investigated as catalyst supports for methane conversion. Examples include Ni/Mg-Al oxide catalysts for steam reforming or dry reforming of methane. The catalysts may be synthesized by constant-pH coprecipitation of metal salts or impregnation of Mg-Al oxides with Ni precursors. Mg-Al double hydroxide nanosheets derived by delamination of Mg-Al layered double hydroxides (LDHs) offer the opportunity to obtain high-surface area supports to achieve high metal dispersion. In this presentation, we report the synthesis of high-surface-area Mg-Al mixed oxides by thermal treatment of Mg-Al LDHs bearing isethionate ($HO(CH)_2SO_3^{â€“}$, Ise). Exchange of interlayer carbonate in MgAl-LDHs bearing $CO_3^{2â€“}$ (MgAl-CO_3) with Ise induces stacking disorder in the lamellar structure as reported in the literature, and allows facile delamination of the resultant material, MgAl-Ise, in water. Despite the highly disordered structure of Ise-containing materials, thermal activation of them in air yielded Mg-Al mixed oxides with low surface area and CO_2 adsorption capacity. In contrast, thermal activation of them in N_2 or N_2 followed by air led to marked increase in CO_2 adsorption capacity up to â‰ˆ800% relative to those obtained by calcination only in air. In addition, these mixed oxides exhibit higher surface area and CO_2 adsorption capacity than those obtained from the initial MgAl-CO_3. We report these new findings and provide the rational of the significant impact of thermal activation conditions on physicochemical properties of Mg-Al mixed oxides. In addition, we report our recent results demonstrating a further improvement in the porous properties of Mg-Al mixed oxides through tuning of the type of interlayer anions and thermal activation conditions.

Fecralloy Partially Oxidizes Methane Selectively

Authors:

Ma, Z., Polytechnique Montreal
Patience, G. S., Polytechnique Montreal
Boffito, D. C., Polytechnique Montreal
Pelegrin, D. C., Polytechnique Montreal

Flaring natural gas contributes to the climate change and air pollution, also wastes a valuable source of energy. Micro Gas-to-Liquids technology (GtL) reduces flared natural gas and carbon emissions while producing value-added synthetic crude - condensate, gasoline, diesel and wax. A mobile micro-refinery unit (MRU), integrating high pressure syngas step (catalytic partial oxidation of methane, CPOX) with Fischer-Tropsch (FT) in a single vessel, reduces investment and operating cost compared to auto thermal reforming and steam reforming.

CPOX catalyst must operate at temperature exceeding 900 °C to achieve 90+% methane conversion and at pressure of 2 MPa to combine with FT process. FeCralloy knitted fibres are ideal supports materials because the oxidized surface alumina layer resists temperature up to 1050 °C . With oxidized FeCralloy as a support, we deposited 0.1%-1% Pt over the fibre surface via solution combustion synthesis. Free of pre-reduction treatment, the selectivity to syngas of the Pt FeCralloy catalyst is superior at both 0.1 MPa and 2 MPa. High pressure restrained the methane conversion and syngas selectivity. Coking deactivates the catalyst, but a quick and simple regeneration step recovers its original reactivity. The FeCralloy surface morphology remains intact after multiple reaction-regeneration cycles.

Greenhouse Gas Footprints of Transportation Fuels Manufactured from Natural Gas Liquids Derived from Shale Gas

Authors:

Chen, Q., The University of Texas at Austin
Dunn, J. B., Northwestern University
Allen, D. T., The University of Texas at Austin

In the processes being developed by National Science Foundation Engineering Research Center on the Innovative and Strategic Transformation of Alkane Resources (CISTAR), natural gas liquids (NGLs) are separated from shale gas, then undergo dehydrogenation and oligomerization to produce transportation fuels. The sale of such unconventional transportation fuels can be limited by the environmental footprints of the fuels in some market segments. Section 526 of the Energy Independence and Security Act (EISA) of 2007[1] requires that alternative transportation fuels sold to the federal government must have lower or equal life cycle greenhouse gas (GHG) emissions compared to petroleum-based fuels. The California Low Carbon Fuel Standard[2] limits the sale of transportation fuels in California based on life cycle GHG emissions. These limitations motivate the understanding of GHG footprints of CISTAR fuels.

Life cycle (well to tank) GHG footprints are a combination of upstream and fuel production components. Upstream emissions of methane (a potent greenhouse gas) and the GHG emissions caused by energy demand of the CISTAR conversion processes are the major contributions to life cycle GHG emissions. The GHG footprints analysis is highly sensitive to the region of origin of the light alkane feedstocks. The fuel production segmentâ€™s contribution to CISTAR fuel life-cycle GHG emissions depends on the details of the conversion scenarios. Preliminary baselines for CISTAR processes show that CISTAR fuel production would be comparable (+/- 10%) to GHG emissions from conventional gasoline production, but more sensitivity analyses in both upstream and fuel production segments need to be done. The results will guide research towards advances that will reduce environmental impacts of CISTAR fuels.

High Efficiency Gas-to-Liquids through Condensation of Small Hydrocarbons Using Corona Discharge Processes

Authors:

Miao, Y., Baylor University

The conversion of small hydrocarbons to longer ones that are liquids at room temperature is the general focus of gas to liquid (GTL) technologies. Recently, the co-authors have been investigating the direct GTL conversion through direct condensation of short hydrocarbons to longer ones using low energy corona discharges.

In this process, low energy electron impact ionization through the use of a non-thermal plasma generated by an atmospheric pressure electric corona discharge is used to generate reactive radicals in the gas phase that subsequently recombine to produce the longer hydrocarbons. By implementing this approach in a microchemical reactor we can take advantage of strong electric field gradients and efficient heat and energy management resulting in the stabilization of the corona discharge in a very non-thermal plasma regime.

Both single-discharge and multi-discharge reactors have been built, and their performance demonstrates the chemical conversion concept presented. A brief investigation of important process parameters shows that the power level per discharge, the discharge gap, and the reactive gas flow rate is particularly important. A high energy efficiency of up to 85% (electrical to chemical energy) can be achieved, and methane conversion can reach up to 25% with high selectivity (80~90%) towards C2+ hydrocarbons.

Ignition-Extinction Analysis of Oxidative Coupling of Methane in Laboratory-Scale Reactors

Authors:

Sun, Z., University of Houston
West, D., SABIC
Balakotaiah, V., University of Houston

A detailed ignition-extinction analysis of laboratory scale reactors with heat exchange with furnace is provided for the case of single and multiple oxidation reactions. It is shown that the same volume or mass of catalyst packed in tubes of different diameter and/or with different lengths of inert sections could lead to different type of ignition-extinction behavior as well as product distribution. The impact of tube diameter, heat exchange time, and length of inert sections on the ignition-extinction behavior is analyzed with either the space time (flow rate) or the furnace temperature as the parameter that is varied experimentally. While the highly exothermic oxidative coupling of methane is used as an example to illustrate the approach, the results are applicable to laboratory reactors in which catalytic partial oxidations, oxidative dehydrogenations or other exothermic reactions occur. The results of ignition-extinction behavior are used to provide guidelines for the interpretation of data and for the scale-up of such processes.

Methane to Products: Integrated Systems Analysis

Authors:

Karwa, S., Shell
Ayoub, P., Shell
Gomez Osorio, M., Shell
Sant, S., Shell

The Methane to products theme of the Long-Range Research program aims to create options for converting methane into valuable products, and doing so with a reduced CO_2 footprint. Currently methane is mainly used as a gas for its heating value, but it can have applications in the chemical industry that are currently based on the cracking of liquid petroleum products. If methane is converted to large scale products that can be sequestered post-consumer use, the CO_2 footprint could be substantially reduced.

This poster explains ways in which methane can be converted into useful products that are of interest to the petrochemicals industry, with the premises of inexpensive renewable energy, and a carbon constrained world.

Natural Gas Role in Future Low Carbon Economies

Authors:

Powell, J. B., Shell International Exploration & Production

How can the abundance of natural gas resulting from recent breakthroughs in shale gas production via fracking technologies, be integrated with renewable energy to reduce carbon footprints for energy and chemicals? The thermal decomposition of methane to produce H_2 and carbon products is a low energy (+76 kjoule/mol), net energy positive means for producing clean fuel and sequestering carbon (Meyer Steinberg et al., Brookhaven labs 1990â€™s). This poster examines use of TDM with help of renewable energy to produce H_2 as clean and green energy carrier, with sequestration of carbon into solid or liquid products (carbon black, nanotubes, other polymers).

Performance of Conical Spouted Beds Technology for Anaerobic Digestion of Wastes from Wine Production

Authors:

San Jose, M. J., University of the Basque Country
Alvarez, S., University of the Basque Country
Lopez, R., University of the Basque Country

Abstract

The wine production generates a large amount of organic wastes, such as skins and scrapes; lees; and vinasse [1], suitable for valorization. The anaerobic digestion produces a methane-rich biogas and and an organic by-product, a digestate, which requires pretreatment for later use for fertilization of fields or as compost. The Spouted Beds technology has been successfully applied for treatment of biomass wastes [2-6].

In this paper, a conical spouted bed has been applied for valorization of grape skins and scrapes wastes by anaerobic digestion. The digestión has been performed with different percentages of substrate and inoculum under temperature in the mesophilic and thermofilic ranges (between 35 and 55 °C), for times between 3 and 6 weeks. The methane content in the generated biogas, measured by the Geotech Biogas-5000 analyzer, was higher than 52%.

Acknowledgements

This work was performed with financial support from the Spanish Ministry of Economy and Competitiveness and co-funded by the European Union through ERDF funds (Projects CTQ2014-59312-P and CTQ2017-89199-P).

References

[1] Wine Institute, OIV report on the world vitivinicultural situation (2017).

[2] San José, M.J., Alvarez, S., Peñas, F.J., García, I. Chem. Eng. Sci. 100 (2013a) 413.

[3] San José, M.J., Alvarez, S., García, Peñas, F.J., Fuel 110 (2013b) 178.

[4] San José, M.J., Alvarez, S., Peñas, F.J., García, I, Chem. Eng. J. 238(15) (2014a) 227.

[5] San José, M.J., Alvarez, S., García, Peñas, F.J., Chem. Eng. Res. Des. 92 (2014b) 672.

[6] San José, M.J., Alvarez, S., López, R., Catal. Today 305 (2018) 13-18.

Small Scale Split Flow Integrated LNG Production Process (SFI-LNG)

Authors:

Ostovar, A., University of Calgary
Nassar, N. N., University of Calgary

A novel split flow integrated LNG production process (SFI-LNG) is developed and optimized in order to significantly reduce power consumption of small-scale LNG production plant which make small-scale LNG plant more efficient and also more environmentally friendly process with lower CO_2 emission.

The main core of SFI-LNG process comprises two fundamental sections: natural gas (NG) purification and NG liquefaction. Compared to a conventional NG liquefaction processes, SFI-LNG process can provide opportunities to economize on number of equipment, for example by using two stages of coil-wound heat exchange instead of three stages and also with lower operation pressure.

No conventional NG treating nor dehydration processes as an objective function is defined in design of NG purification section. In order to prevent the CO_2 freezing in the NG liquefaction exchangers, CO_2 content must be reduced from 3mol% to less than 50ppm. Part of the invention cryogenic CO_2 removal process has developed and can purify dried natural gas at cryogenic temperature by utilizing a new nanoparticle packed column as a physical adsorption bed.

The power consumption as an objective function is optimized with key parameters and the optimization results show that the NG liquefaction section can provide for significant reductions in total power consumption by up to 13.2% compared to conventional technologies. This in turn, the mass flow of mixed refrigerant cycle (MR) can be reduced compared to conventional processes by up to 16.1%. In particular, in conventional propane-mixed refrigerant cycle (C3-MR) processes, pre-chilling by propane is typically required, however, in SFI-LNG process, no pre-chilling is required and the mass flow of propane refrigerant loop can be smaller than in conventional installations by up to 18.8%.

AUTHOR INDEX